laserpronet.com
Empowering the Laser Workforce

What we do at LaserPronet

- Screening Exams
- Professional Development Courses
- Professional Growth Plans
- Certifications

1

Solid State Laser and their Commin Problems: Laser Resonator Wavelength Selection

A. Laser Technician Series

Laser Technician I Job Description: Level I Technicians perform final tests on lasers and laser systems to ensure that they fully comply with customer specifications. All tests are completely documented for both internal and external use.

Environment: Test Laboratory.

Requirements: Must understand laser and optics principles and, laser beam performance specifications. Experience testing/evaluating laser beams and working with photonics test equipment is a must. Also, ability to follow prescribed written procedures, directions and strict adherence to best laser lab and manufacturing practices are required.

Minimum Educational Requirements: Candidate must hold a certificate/degree in Laser/Electro-Optics Technology or related discipline

Laser Technician II Job Description: Level II Technicians assemble and troubleshoot common optical problems in laser systems. They perform final tests on the systems to ensure that they fully comply with customer specifications and all tests are completely documented for both internal and external use.

Environment: Manufacturing and Test Lab

Requirements: Must understand Wave and Geometrical Optics, Gaussian Beam Propagation, Nonlinear Optics and how acousto- and electro-optics modulators, optical components and accessories work. Experience aligning laser systems, troubleshooting laser beam aberrations and working with photonics test equipment is a must. Also, ability to follow prescribed written procedures, directions and strict adherence to best laser lab and manufacturing practices are required.

Minimum Educational Requirements: Candidate must hold a certificate/degree in Laser/Electro-Optics Technology or related discipline.

Laser Technician III Job Description: Level III Technicians assemble, align, burn-in, test, and tune/troubleshoot laser heads until they meet all performance specifications.

Environment: Manufacturing

Requirements: Must understand the fundamentals of solid state laser technology, accessories, and support systems. Experience aligning laser systems and cavities/resonators and using photonics test equipment is a must. Also, ability to follow prescribed written procedures, directions and strict adherence to best laser lab and manufacturing practices are required.

Minimum Educational Requirements: Candidate must hold a certificate/degree in Laser/Electro-Optics Technology or related discipline

Level 3 Volumes - Solid State Lasers and their Common Problems
Volume 1: Solid State Laser Optical Pumps
Volume 2: Amplifying Crystals
Volume 3: Laser Resonator Wavelength Selection
Volume 4: Laser Resonator Transverse Modes

Technician IV Job Description: Level IV Technicians support Research and Development (R&D) and, customers. Customer Support/Technical Service Technicians work on deployed lasers and laser systems in support of existing customers. R & D/Engineering Technicians support scientists and engineers improve existing, and also create the next generation laser technologies. Level IV Technicians work under limited supervision.

Environment: Research/Engineering Lab and Field

Requirements: Must have a thorough understanding of solid state laser technologies and support systems, experimentation and research protocols. Experience collecting and analyzing data, troubleshooting/problem-solving, using MS Office to generate test reports and writing technical reports is required. Customer

Support/Service technicians may have to travel to customer sites. **Minimum Educational Requirements:** Candidate must hold a certificate/degree in Laser/Electro-Optics Technology or related discipline.

B. Introduction

A monochromatic laser beam has one wavelength and that is the general goal of almost all lasers except for broadband/ultrafst lasers.

Monochromatic laser beams are achieved through the preservation of one wavelnegth while annihilating the rest or any available. Common ways to accomplish this is the use of (1) a laser cavity length that would only fit /accommodate an interal number of wavelengths, (2) insertion of optical coatings on laser mirrors and other intra-cavity components, and (3) insertion of etalons within a laser resonant cavity

C. Optical Resonator/Resonant Cavity Longitudinal Modes

- A resonant cavity/resonator is composed of at least two aligned mirrors
- In an optical resonator mirrors are needed for directionality **only**

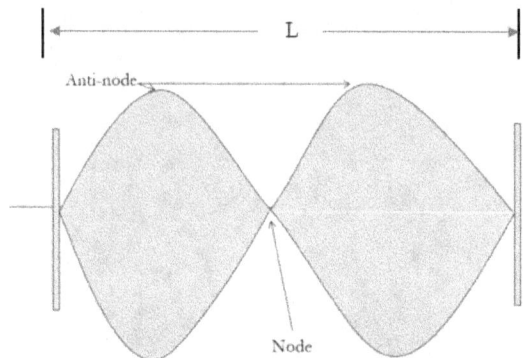

- In an optical resonant cavity of length L, the incident beam will interfere constructively such that the exiting waves are in phase if L is equal to an integral, n, number of half wavelengths, λ,

    $L=n\lambda/2$ where n=1,2,3,4…...

  - Any wave that meets the afore mentioned condition will resonate in the cavity i.e. resonant cavity

- Any wave that resonates in a cavity will exhibit a standing wave pattern
- Note that the cavity length, L, determines the number of longitudinal modes, n, a resonant cavity can support given a wavelength, λ.

- Standing waves are formed in a cavity only for wavelengths for which have nodes at the end cavity.

- Each standing wave in a resonator/resonant cavity represents a longitudinal/axial mode
- A vibrating guitar string exhibits standing wave pattern
- Longitudinal/axial modes in a resonator are separated in frequency space,

$$\Delta f = \frac{c}{2nl}$$

Where
c is the speed of electromagnetic waves/light, and
L is the length of the resonator
n is the refractive index along the optical path within the resonator

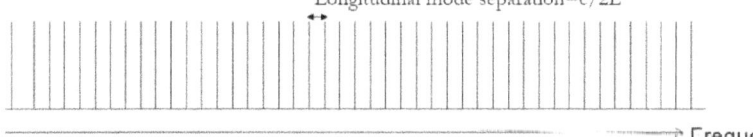

- The longer the resonator, L, the more clustered are the longitudinal modes of the resonator
- A Fabry-Perot interferometer is an example of a resonator without an amplifier

D. Optics Coatings

The Reflectance, R, at normal incidence of ray of light in medium 1 (n_1) onto a transparent material with an index of refraction n_2 is given by

$$R=(n_1-n_2)^2/(n_1+n_2)^2$$

If for example the incidence medium is air (n_1=1) and the reflecting medium is glass (n_2=1.5) it can be shown that 4% of the incident light is reflected at the air-glass interface.

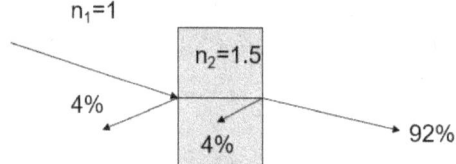

It can also be shown that 4% of the incident light is also reflected at the glass-air interface.

The Reflectance of an optical surface can be altered to

1. reduce losses by transmitting optics
2. enhance both the transmission and reflection light and
3. stop the reflected-on light

through the deposition of optical coatings. A common optical coating material is magnesium fluoride (MgF_2)

Reduction of Transmission Losses by Optical Coatings

Rule for Assigning Indices of Refraction $n_3>n_2>n_1$

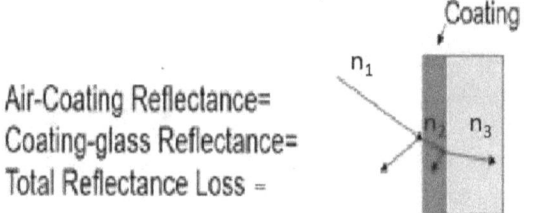

Air-Coating Reflectance=
Coating-glass Reflectance=
Total Reflectance Loss =

Example 1: Assume a planar piece of glass is deposited with an optical coating of n=1.25

1	Calculate the percentage of incident light reflected at the air-coating interface	1.23
2	Calculate the percentage of incident light reflected at the coating-glass interface *(100-1.23)*.83*	.82
3	% Reflection by bare glass=	4
4	Total % reflection by coated optic (1+2)	2.05
5	% Reflection reduction (3-4)	1.95

.

Example 2: Assume a planar piece of glass is deposited with an optical coating of n=1.2

1	Calculate the percentage of incident light reflected at the air-coating interface	.83
2	Calculate the percentage of incident light reflected at the coating-glass interface (100-.83)*99.17%	1.22
3	% Reflection by bare glass=	4
4	Total % reflection by coated optic (1+2)	2.05
5	% Reflection reduction (3-4)	1.95

Example 3: Assume a planar piece of glass is deposited with an optical coating of n=1.3

1	Calculate the percentage of incident light reflected at the air-coating interface	1.7
2	Calculate the percentage of incident light reflected at the coating-glass interface	.50
3	% Reflection by bare glass=	4
4	Total % reflection by coated optic (1+2)	2.2
5	% Reflection reduction (3-4)	1.8

Make a general statement as to how the reduction of reflected light changes as the index of refraction of the optical coatings deposited on the glass is increased /decreased.

_____

_____

_____

_____

Enhancement and Elimination of Light Reflection by Optical Coatings
Optical coatings work on the principle of Interference

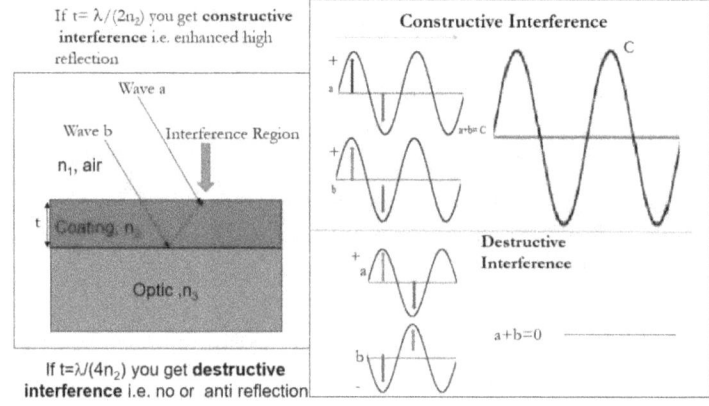

- In an optical coating light reflected from one layer can be made to interfere with light constructively/destructively
    - In an optical coating of index of refraction n when coherent light of wavelength λ is transmitted through and then reflected into the coating layer it will interfere constructively with incident light at the exit edge of the coating if the coating thickness (t =) $\lambda/2n$
        - This enhances reflections at the surface of the optics
    - In an optical coating of index of refraction n when coherent light of wavelength λ is transmitted through and then reflected into the coating layer it will

interfere destructively with incident light at the exit edge of the coating if the coating thickness (t=) $\lambda/4n$

- This eliminates light reflection i.e. anti-reflection (AR) coating

E. Laser Resonators/Resonant Cavities and Wavelength Selection

Laser Mirrors and Wavelength Selection

- In a laser resonator, the general goal is to oscillate, amplify and outlet one wavelength
- Coated optics/mirrors are used for wavelength selection and reflecting the selected wavelength to the amplifier in a resonant cavity
- Optical coatings on mirrors are used to select wavelengths to
  - eliminate,
  - oscillate and amplify and
  - then output from a resonant cavity through the OC

- The Output Coupler (OC) partially reflects inwards transmits outward the wavelength of interest
  - The Output Coupler (OC) will
    - destroy wavelengths that are not of interest via coatings
    - keep the wavelength(s) of interest oscillating in a laser cavity
    - For example
      - If a laser cavity has 100W within it then its output power would be 4 Watts, if it has 4% OC transmission.
      - If a 100-Watt laser cavity has an OC with a Reflectivity, R, of 96% the laser will output 4 Watts.
      - If a HeNe laser, λ=632.8 nm, with an Output Coupler (OC) with a reflectivity of 95% will transmit (T) 5% of power of wavelength of interest.
- The High Reflector (HR) 100% inwards towards the amplify the wavelength of interest

14

- High Reflectors, HR, are nominally assigned reflectivity values of 100% on any lasers. Experimental findings show that the reflectivity of HRs is slight less than 100 i.e. 99.xxx %

- In general T+R+A=100%
    - Where A is the absorbance and we assume its zero (A=0) or negligible for all optics under analysis, so the above equation can be recast as
        - T+R=100%
        - If we can therefore measure a laser mirror's transmission, T, at a specific wavelength we can go on to calculate its reflectivity, R, as follows
        - 100-T=R
            - Note that generally reflectivity, R, is specified on optics while some spectrometers may only measure transmission, T, therefore it is left to the user to convert transmissions to reflectivity's.

F. Etalons and Laser Wavelength Selection

- Etalons are based on the Fabry-Perot resonator concept.
  - $L_{etalon}=n\lambda/2$

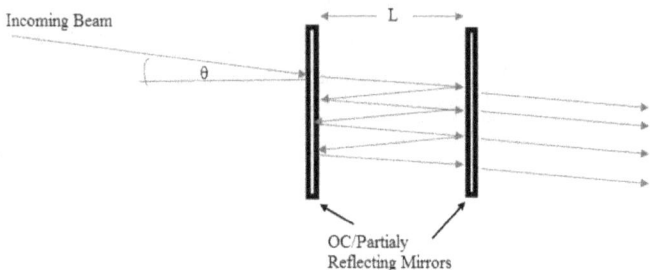

θ is generally zero but has been tilted in the diagram for illustration and lucidity purposes

- They are made with two parallel surfaces and in between there it could be solid material or air-filled.
- They are inserted in laser resonant cavities to narrow-down/limit a laser's emitted output wavelength
- Additional wavelength selectivity in a laser resonator/cavity can therefore be achieved through the use etalons
  - Etalons are not a substitute for HR and OC.

G. Laser Output Wavelength

- Each supported longitudinal mode in a laser cavity represents potential   laser output wavelength

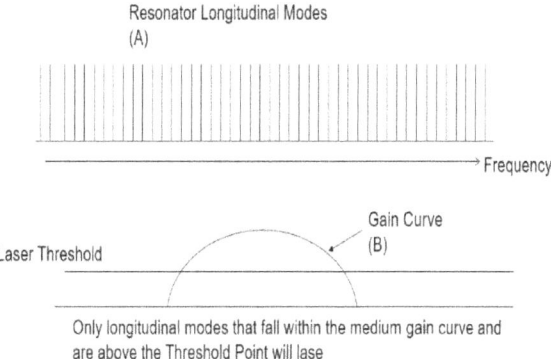

Resonator Longitudinal Modes
(A)

Frequency

Laser Threshold

Gain Curve
(B)

Only longitudinal modes that fall within the medium gain curve and
are above the Threshold Point will lase

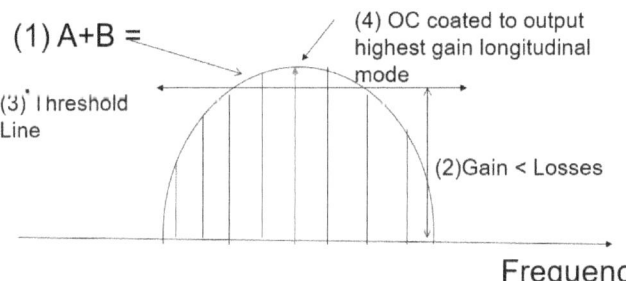

(1) A+B =

(4) OC coated to output
highest gain longitudinal
mode

(3) Threshold
Line

(2)Gain < Losses

Frequency

Wavelengths outside gain curve (A+B) and
below (3) Threshold are excluded from lasing.
In addition OC coatings (4) will output
selected longitudinal mode and annihilate
all others if present

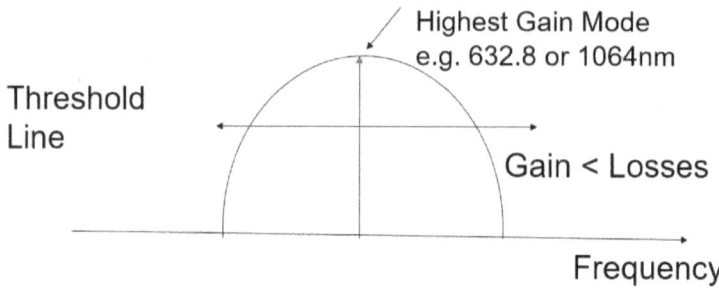

Threshold
Line

Highest Gain Mode
e.g. 632.8 or 1064nm

Gain < Losses

Frequency

•Optical coatings on OC eliminate none
desired wavelengths above Threshold.
•In addition, other devices such as etalons can be inserted in the cavity
to eliminate undesired wavelengths.

- A laser's output wavelength is determined by length of the cavity, Gain Curve of the active medium and HR and OC optical coatings
- A laser's output wavelength range can additionally be restricted by the insertion of an etalon in the resonator

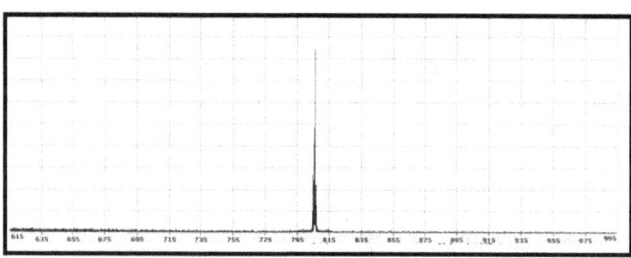

H. Laser Resonator Wavelength Selection
Self-Test Questions

A. Optics Coatings
B. Optical Resonators/Resonant Cavity Longitudinal Modes
C. Laser Resonators/Resonant Cavities and Wavelength Selection
D. Etalons and Laser Wavelength Selection
E. Laser Output Wavelength

A. Optical Resonator/Resonant Cavity Longitudinal Modes

1. A resonant cavity/resonator is composed of at least _____
    aligned mirrors
    a. two
    b. three
    c. four
    d. Any of the above
    e. None of the above

2. In an optical resonator mirrors are needed for _____.
    a. directionality
    b. amplification
    c. a and b
    d. None of the above

3. In an optical resonant cavity of length L, the incident beam
    will interfere constructively such that the waves are in phase
    if L is equal to an integral, n, number of
    _____wavelengths, λ.
    a. half
    b. full
    c. quarter
    d. all the above
    e. None of the above

4. Any wave that resonates in a cavity will exhibit a _____
    wave pattern
    a. standing
    b. spherical
    c. Gaussian
    d. Any of the above
    e. None of the above

5. Resonator cavity length, L, determines the number of
_____ modes it can support given a wavelength, λ
   a. longitudinal
   b. axial
   c. transverse
   d. a and b
   e. None of the above

6. A standing wave has maximal energy at the _____.
   a. notes
   b. nodes
   c. antinodes
   d. Any of the above
   e. None of the above

7. A standing wave exhibits no energy at the _____.
   a. notes
   b. nodes
   c. antinodes
   d. Any of the above
   e. None of the above

8. Standing waves are formed only for wavelengths that form
_____ at the resonator ends/mirrors.
   a. nodes
   b. antinodes
   c. beats
   d. Any of the above
   e. None of the above

9. Longitudinal/axial modes in a resonator are separated in frequency spaceΔf, by _____ spacing.
   a. $\Delta f = c/2nL$
   b. $\Delta f = L/2nc$
   c. $\Delta f = n/2cL$
   d. Any of the above
   e. None of the above

   Where
   c is the speed of electromagnetic waves/light, and
   L is the length of the resonator
   n is the refractive index along the optical path within the resonator

10. The longer the resonator, L, the _____ clustered are the longitudinal/axial modes in a resonator
    a. more
    b. less
    c. (Has no impact on the clustering/density of modes)

11. A _____ interferometer is an example of an optical resonator.
    a. Fabry-Perot
    b. Michelson
    c. a and b
    d. None of the above

B. Optics Coatings

12. The Reflectance, R, at normal incidence of ray of light in medium 1 (n_1) onto a transparent material with an index of refraction n_2 is given by _____.
    a. $R=(n_1-n_2)^2/(n_1+n_2)^2$
    b. $R=(n_2-n_1)^2/(n_2+n_1)^2$
    c. $R=(n_1+n_2)^2/(n_1-n_2)^2$
    d. a and b
    e. None of the above

13. If the incidence medium is air ($n_1=1$) and a reflecting crown glass block has an index of refraction n_2 of 1.6 ($n_2=1.6$) then the Reflectance, R (%), at the air-crown glass interface is

    _____.
    a. R= 4%
    b. R= 5.3%
    c. R= 7%
    d. Any of the above
    e. Nonene the above

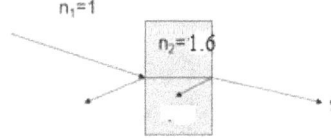

14. As light exits a crown glass block (n=1.6) into air ($n_1=1$) the Reflectance, R (%), at the crown glass-air interface
    a. R= 3.5 %
    b. R= 5.3%
    c. R= 4 %
    d. Any of the above
    e. Nonene the above

15. The Reflectance of a transparent optical surface can be altered to
    a. reduce transmission losses and enhance transmission
    b. reduce reflection losses
    c. stop the reflection
    d. all the above

16. Optical coatings work on the principle of _____.
    a. Interference
    b. Diffraction
    c. Polarization
    d. a and b
    e. None of the above

17. In an optical coating light reflected from one layer interface can be made to interfere with light reflected from another layer interface _____.
    a. constructively
    b. destructively
    c. a and b
    d. None of the above

18. In an optical coating of index of refraction n when coherent light of wavelength λ is transmitted through it and then reflected into the coating layer it will interfere constructively with incident light at the exit edge of the coating if the coating thickness t is _____. Assume the light beams to be in-phase before encountering the coatings/optics.
    a. $t = \lambda/2n$
    b. $t = \lambda/4n$
    c. any of the above
    d. None of the above

19. In an optical coating of index of refraction n when coherent light of wavelength λ is transmitted through it and then reflected into the coating layer it will interfere destructively with incident light at the exit edge of the coating if the coating thickness t is _____. Assume the light beams to be in-phase before encountering the coated optics.

    a. t = λ/2n

    b. t = λ/4n

    c. any of the above

    d. None of the above

20. In an optical coating of index of refraction n when coherent light of wavelength λ is transmitted through and then reflected into the coating layer and interferes constructively with incident light at the exit edge of the coating, this _____.

    a. enhances reflections at the surface of the optics

    b. eliminates light reflection

    c. a or b

    d. None of the above

21. In an optical coating of index of refraction n when coherent light of wavelength λ is transmitted through and then reflected into the coating layer and interferes destructively with incident light at the exit edge of the coating, this _____.

    a. enhances reflections at the surface of the optics

    b. eliminates light reflection

    c. a or b

    d. None of the above

C. Laser Resonators/Resonant Cavities and Wavelength Selection

Laser Mirrors and Wavelength Selection

22. In a laser resonator, the general goal is to _____ one wavelength
    a. oscillate
    b. amplify
    c. output
    d. all the above
    e. None of the above

23. Optical coatings on laser mirrors are used to select wavelengths to _____
    a. eliminate,
    b. oscillate and amplify and
    c. then output from a resonant cavity through the OC
    d. all the above
    e. None of the above

24. In a laser resonator mirrors are needed for
    a. directionality
    b. amplification
    c. a and b
    d. None of the above

25. The Output Coupler (OC) will
    a. destroy wavelengths that are not of interest via coatings
    b. keep the wavelength(s) of interest oscillating in a laser cavity
    c. a and b
    d. None of the above

26. If a laser resonant cavity has 100W within then its output power would be _____ Watts if it has 6% OC transmission.
    a. 6
    b. 12
    c. 94
    d. Any of the above
    e. None of the above

27. If a 400-Watt laser cavity has an OC with a Reflectivity, R, of 96% the laser will output _____ Watts.
    a. 4
    b. 40
    c. 16
    d. Any of the above
    e. None of the above

28. If a HeNe laser resonator with an Output Coupler (OC) with a reflectivity of 95% will transmit (T) _____ % of power of wavelength of interest.
    a. 95
    b. 5
    c. 4
    d. Any of the above
    e. None of the above

Transmission
Spectrograph **A**

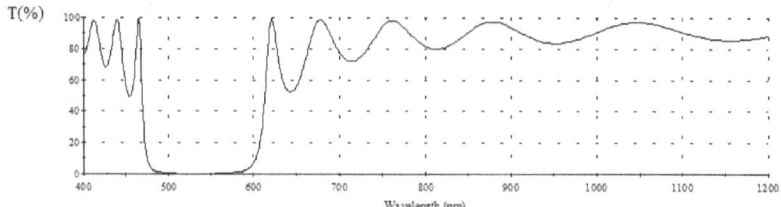

Transmission
Spectrograph **B**

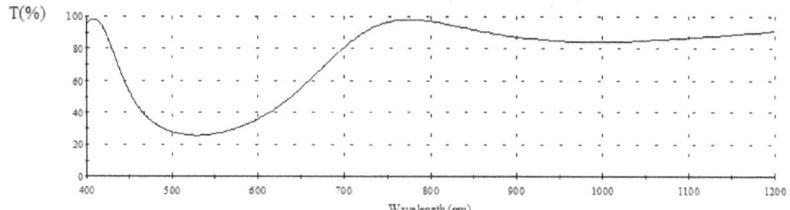

29. Spectrum above _____ is that of HR for a laser emitting 532 nm

    a. A

    b. B

    c. A or B

    d. None of the above

30. Spectrum above _____ is that of OC for a laser emitting 532 nm

    a. A

    b. B

    c. A or B

    d. None of the above

31. Optic A approximately has _____ Percent Transmission (%) at 532 nm
    a. 0
    b. 100
    c. 50
    d. Any of the above
    e. None of the above

32. Optic A approximately has _____ Percent Reflectivity (%) at 532 nm
    a. 0
    b. 100
    c. 50
    d. None of the above

33. Optic B approximately has _____ Percent Transmission (%) at 532 nm
    a. 20
    b. 30
    c. 25
    d. None of the above

34. Optic B approximately has _____ Percent Reflectivity (%) at 532 nm
    a. 80
    b. 70
    c. 75
    d. None of the above

D. Etalons and Laser Wavelength Selection

35. Etalons are based on the _____ resonator concept.
    a. Fabry-Perot
    b. Michelson
    c. a and b
    d. None of the above

36. Etalons are inserted in laser resonant cavities to _____ a laser's emitted output wavelengths
    a. narrow-down/limit
    b. expand
    c. a and b
    d. None of the above

37. Etalons can _____ used as a substitute for HR and/or OC.
    a. be
    b. not be

E. Laser Output Wavelength

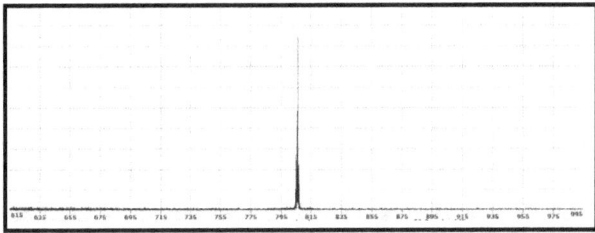

Figure: Wavelength (nm) output of a laser spectrum analyze

38. The spectrum analyzer is showing an output of about
    _____ nm
    a. 795
    b. 815
    c. 810
    d. Any of the above

39. If the monochromatic laser being sampled output and show more than one longitudinal mode line on the spectrum analyzer, additional lines are due to faulty _____.
    a. optical coatings on OC
    b. optical coatings on HR
    c. a and b
    d. all the above
    e. None of the above

40. A laser's output wavelength is also one of its _____ modes in the resonant cavity
    a. axial
    b. longitudinal
    c. transverse
    d. a and b
    e. None of the above

31

www.ingramcontent.com/pod-product-compliance
Lightning Source LLC
Chambersburg PA
CBHW030041230526
45472CB00002B/613